NATIONAL GEOGRAPHIC

Ladders

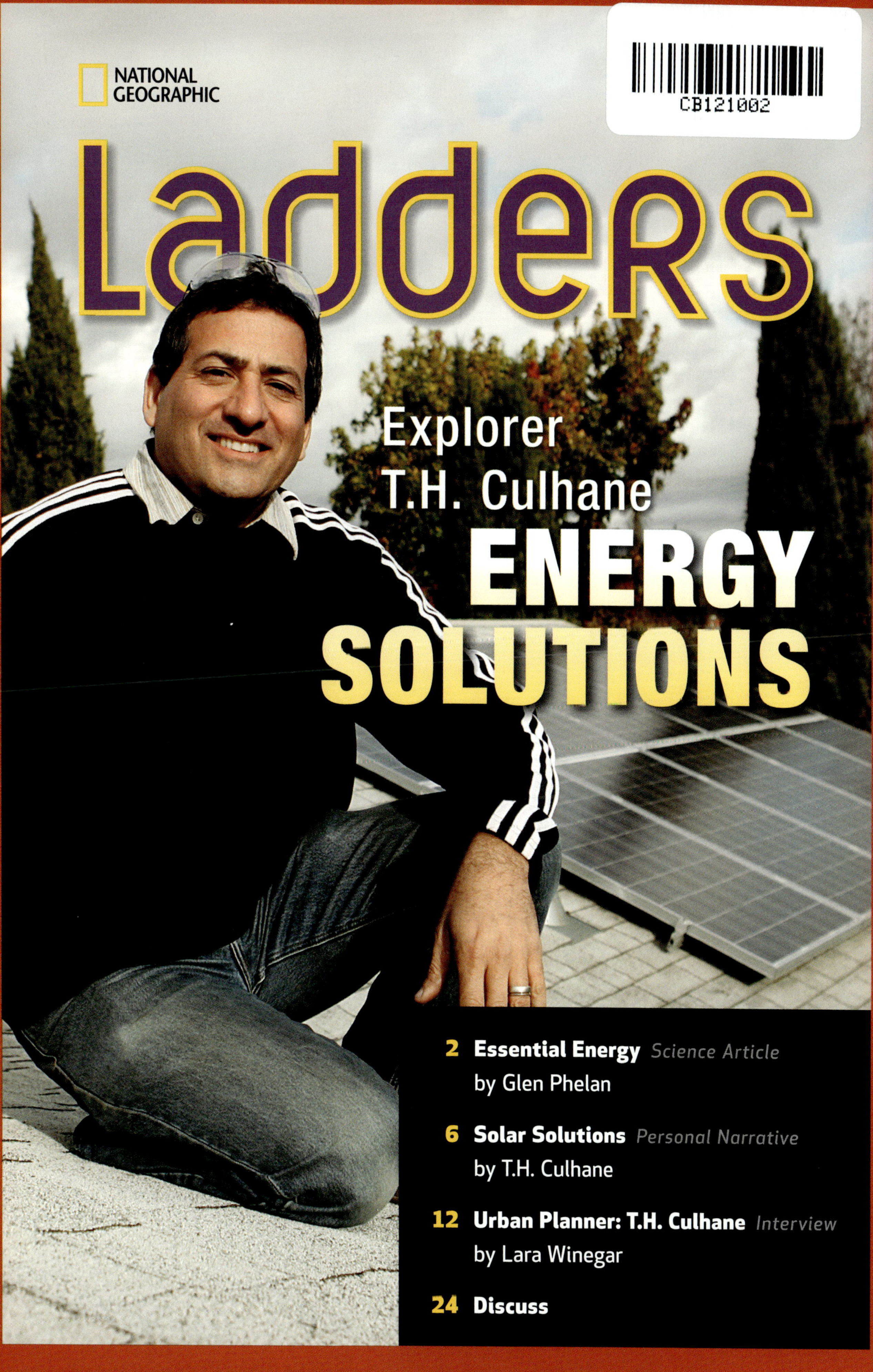

Explorer
T.H. Culhane

ENERGY SOLUTIONS

GENRE Science Article

Read to find out how people in different parts of the world get energy.

ESSENTIAL Energy

by Glen Phelan

"We all need energy. What's the best way to get it? I found some answers where I least expected them—in villages deep in the rain forest." –T.H. Culhane

Could you go a day without using any convenient sources of energy? That's only 24 hours—no problem, right? Just remember—almost everything you do uses energy.

So, where do we get most of the energy we use? Much of it comes from fossil fuels—oil, coal, and natural gas. Power plants burn these energy sources to produce electricity. Then, a giant network, or grid, of wires carries electricity across the United States. The grid connects cities, towns, and farms to power plants. A network of pipes also provides natural gas to many cities and towns.

These energy sources aren't perfect though. There's a limited amount of fossil fuel, and burning it causes pollution. That goes for gasoline, too. It comes from oil.

THOMAS TAHA RASSAM CULHANE, or **T.H.,** is an urban planner. Unlike most urban planners, T.H. doesn't lay out streets or parks. He helps people in cities find ways to solve challenging problems—energy problems and garbage problems.

People in many places do not have electricity or natural gas. This woman in an African village cooks over an open fire.

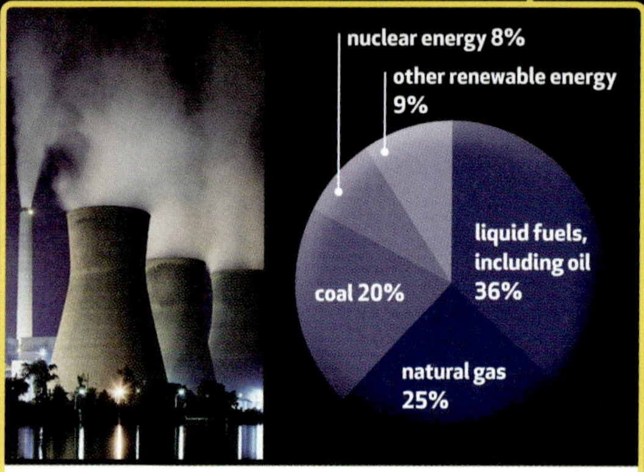

Most energy in the U.S. comes from burning fossil fuels. Some electricity comes from **nuclear energy.** Nuclear energy comes from splitting atoms. Atoms are tiny particles that make up matter. Only a small part of U.S. energy comes from renewable sources such as solar and wind energy. These sources can be used over and over without running out.

Much of Brazil's electricity comes from water. The force of moving water spins a generator to produce electricity. This is **hydroelectric energy** and it is renewable.

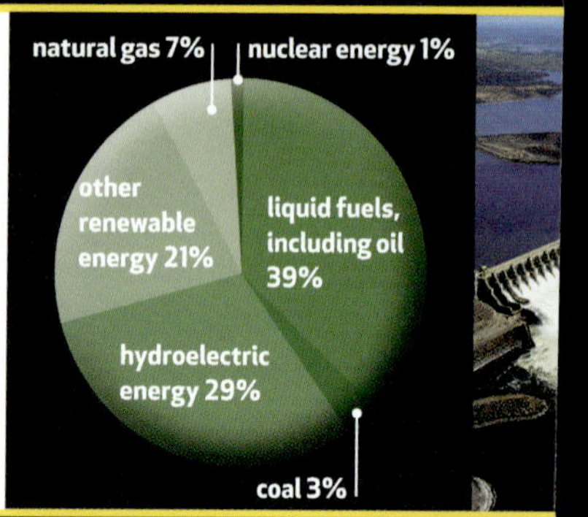

Energy Sources AROUND THE WORLD

For you, getting energy is as easy as flipping a switch. But many places in the world cannot easily get energy. Remote areas of the world often are not part of an electric grid. Take a look at this world map. The map was made from many nighttime photos taken

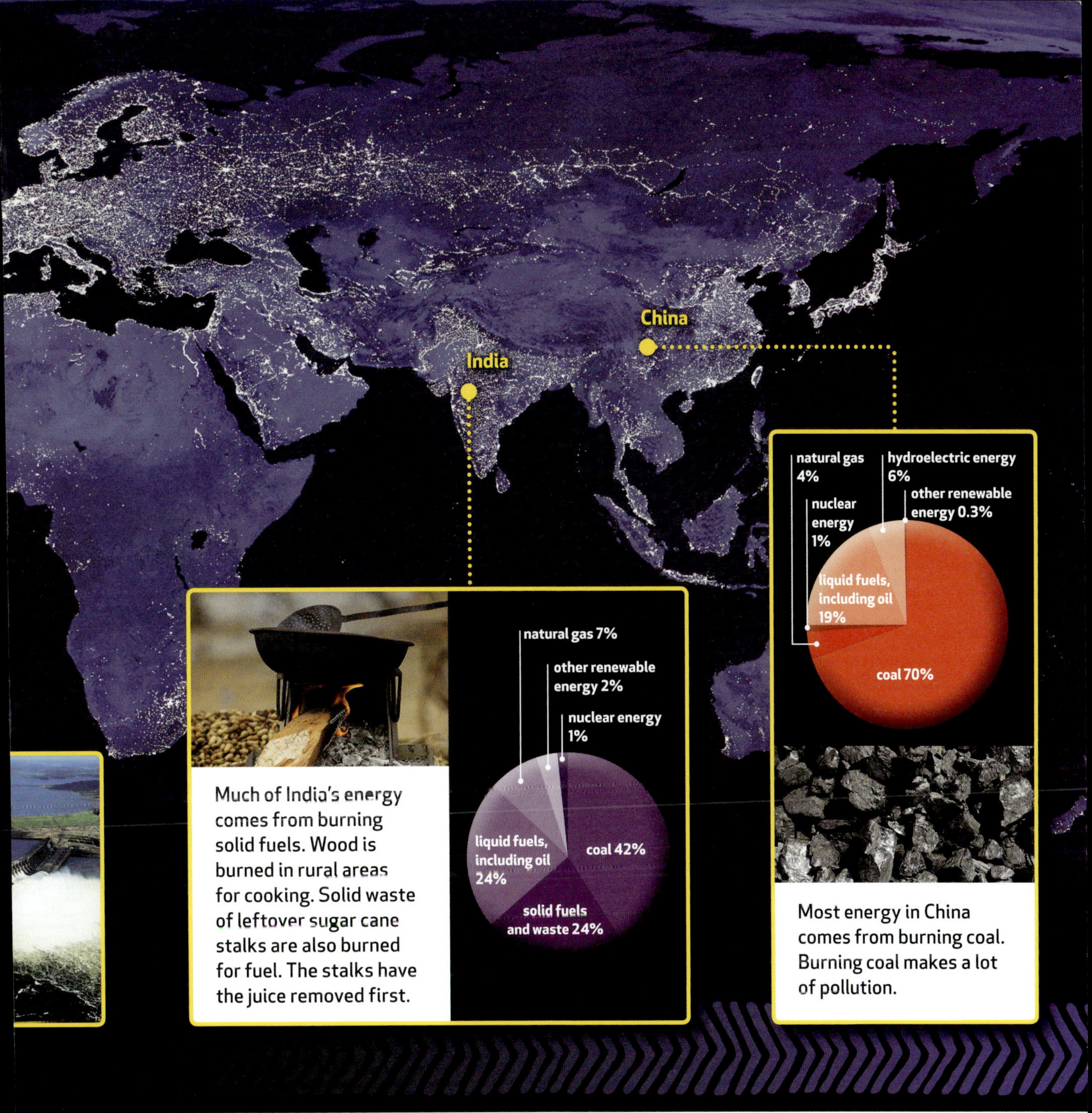

Much of India's energy comes from burning solid fuels. Wood is burned in rural areas for cooking. Solid waste of leftover sugar cane stalks are also burned for fuel. The stalks have the juice removed first.

Most energy in China comes from burning coal. Burning coal makes a lot of pollution.

from space. See where electricity lights up our world—and where it doesn't. The graphs give more information on energy use.

In several of the places with few or no lights, T.H. Culhane has helped people find creative ways to get energy—without burning anything. Now that's a real challenge.

Check In What kinds of energy are used in China?

GENRE Personal Narrative

Read to find out about using solar energy in the mountains of Nepal.

Solar Solutions

by T.H. Culhane

It's a sunny day in the valley. The day is bright, but the sunlight doesn't warm things up much here at 4,400 meters (14,436 feet). Today my team is in Dingboche, a small village in the Himalaya in Nepal. Towering peaks rise all around us. In fact, the tallest mountain in the world—Mount Everest—is within sight. Talk about being off the grid!

About 200 people live in this tiny village. But each summer, hundreds of visitors wander through on their way to climb Mount Everest. And all of these people need energy. They burn fuel to keep warm and cook food. Some of that fuel is kerosene, a liquid made from oil. Some is natural gas. Yaks haul tanks of these fuels up the mountain. Another fuel is the juniper shrub. These shrubs help hold soil in place, which is important in the mountains. But some hillsides that used to be covered with juniper are now bare. People have been burning the shrubs for firewood.

Hauling in tanks of fuel is difficult. And burning the junipers damages the environment. I have a better idea for getting energy, and it's right here in the mountains—sunlight!

A yak carries a tank of natural gas up the Himalaya.

Some people burn juniper shrubs for heat. How else can people get energy in the mountains?

Sunlight may not warm the air much up here, but it sure can heat up the air inside these special glass tubes. On this trip, we are installing these tubes on the roof of a building. Let's climb up and start using this energy.

We're going to start by using sunlight to heat some water. The walls of each of these special solar tubes have two layers of glass with an empty space in between. The outer layer allows sunlight to pass through it. Sunlight is a form of **solar energy**. When the sunlight strikes the inner layer of glass, it is changed into heat energy. The heat energy causes the inside of the tubes to heat up.

T.H. slides glass tubes into place on a water tank. These glass tubes collect the solar energy that heats water in the tank.

Think of the way sunlight shines through car windows and heats up the inside of the car. The seats in the car can get very hot!

There is a liquid in the inside tubes. This liquid becomes hot as the tubes absorb solar energy. The solar tubes are attached to a water tank and these tubes heat the water in the tank. One of these tubes won't heat a lot of water. But 20 or 30 of them can heat a tank of water to 80°C (176°F). That's plenty hot for washing and cooking.

OK, the last tube is in place. Let's walk to the other section of the roof and build another "bright idea!"

T.H. and his team use the talents of the local people. They find solutions to their energy needs. Most of the work is simpler than work local tradesmen do.

Using sunlight to heat water is a perfect way to use the sun's energy. If only we could use sunlight to produce electricity. Well, we can! Our team is installing panels of **solar cells** that do exactly that. A solar cell is made of special materials. The materials produce an electric current when light hits them. With 72 solar cells, you can run radios or a laptop computer. If hundreds or thousands of solar cells are connected into solar panels, you can run microwaves, refrigerators, or big-screen televisions.

Our team installs wires to carry energy from the solar panels.

Solar energy is nothing new. In fact, almost all the energy we use comes from the sun. Coal comes from plants that used sunlight to grow millions of years ago. Oil and natural gas come from tiny ancient animals that ate tinier living things that used sunlight to grow. Sunlight heats Earth. The heat makes wind blow and moves water through the environment. So energy from wind and moving water really come from solar energy. Using these ideas, I help teach people how to put solar energy to work. It just makes sense!

Solar energy helps people in Nepal meet their energy needs.

Check In How are the solar hot water system and the solar panels good energy solutions in remote places such as Dingboche?

GENRE Interview

Read to find out how T.H. Culhane uses engineering solutions to turn garbage into usable energy.

URBAN PLANNER: T.H. Culhane

by Lara Winegar

What do you do when you wake up in the morning? Do you turn on the lights? Maybe you put bread in the toaster for breakfast. However you start your day, chances are you are using electricity.

Over 1.5 billion people in the world do not have reliable access to electricity. T.H. Culhane is an urban planner. He is teaching people how to meet their energy needs.

T.H. Culhane is an expert at finding solutions to energy problems. T.H. has helped people all over the world. He has solved problems in mountain villages in Nepal. He has also helped find solutions in Cairo, Egypt.

T.H. got his start in the rain forests of Borneo. He studied people in the rain forest. They thrived in their environment. He thought, "How can we do the same thing in cities?"

Let's talk with T.H. about some of his energy solutions.

T.H. Culhane and a friend are checking out dinner. Pork is cooking in this experimental stove. The stove does not burn trees from the forest. It uses sawdust waste instead.

NATIONAL GEOGRAPHIC: Why did you go to rain forests?

T.H. CULHANE: I learned that people who live in the rain forest get all of their energy from the forest. I went to live with them to learn how they did it.

NG: Your work takes you all over the world. You always try to find sustainable energy solutions. What does that mean?

T.H. CULHANE: *Sustainable* means doing something in such a way that you can keep doing it. Here's an example. You can save some of your garden seeds from this year. Then plant these seeds next year. You can do this year after year. You'll never run out of seeds. That's a sustainable way to grow a garden.

"THE WORLD NEEDS ENERGY SOLUTIONS NOW." T.H. CULHANE

NG: You work in some of the poorest parts of the world. Can the people who live in these places afford these sustainable systems?

T.H. CULHANE: Yes, indeed they can! Poor communities are rich with possibilities and innovative ideas. These sustainable systems can be built using mostly recycled materials. They cost very little and we provide the basic plans. Then the people work hard to help themselves. They use their knowledge and creativity to adjust the plans and help build the systems.

People in Tanzania, Africa, find creative energy solutions.

Identifying and Describing the Problem

NATIONAL GEOGRAPHIC: You were educated as an urban planner. But you also sound like an **engineer.** Is engineering important in your work?

T.H. CULHANE: Yes, it is. I didn't study engineering in school, but I approach problems like an engineer. You probably do that, too. Think about what happens if you get a basketball stuck between the rim and the backboard. First, you identify the problem, the basketball is stuck. Then, you use what you know to figure out how to get the ball down by hitting it with another ball. That is solving the problem like an engineer.

NG: What energy problems have you identified in the areas where you have worked?

T.H. Culhane explains some of the benefits of the biodigester to village leaders.

"I DIDN'T STUDY ENGINEERING IN SCHOOL, BUT I APPROACH PROBLEMS LIKE AN ENGINEER."

T.H. CULHANE: One, how can we get enough energy? Most energy humans use originally comes from the sun. We have to transfer energy from the sun into a form we can use. Two, how can we get rid of wastes without harming the environment?

NG: How do we transfer energy into forms we can use?

T.H. CULHANE: Solar energy is a sustainable solution to an energy problem. It doesn't run out. We have developed a device called a **biodigester.** The biodigester can transfer energy in waste into energy we can use.

A villager sets the foundation for the bottom of the biodigester.

Engineering Solutions

NATIONAL GEOGRAPHIC: How did you get the idea of turning garbage into energy?

T.H. CULHANE: Getting rid of wastes without harming the environment is important. Some cultures understand that everything has a use. Many people collect plastic, glass, and metal that others throw out. It's valuable stuff just waiting for someone to figure out how to use it. You can build solar water heaters from recycled materials. Food wastes and toilet wastes are also useful. They are forms of energy. So a problem is also a solution!

^ Joram Samson is working on the holding tank for a biodigester.

Joram Samson is a local trainer for the biodigester. He teaches his neighbors how the system works.

NG: How do you get useful energy from wastes?

T.H. CULHANE: First we build a holding tank. Often we use recycled materials that others have thrown out. Then we grind up our food scraps in a garbage disposal. The chopped up food gets piped to the biodigester. So does human waste from the toilet. Then bacteria feed on the wastes in the biodigester. As they feed, they produce methane gas. We burn the methane gas on the stove for cooking. You can even burn the gas in a generator to produce electricity.

NG: Does the biodigester produce anything besides gas?

T.H. CULHANE: Yes, the process also produces a rich liquid fertilizer. We use it in our garden.

Cow manure is added to the mix. Straw that was mixed with the manure is then removed. Straw can block gas from moving up pipes.

Designing and Improving Solutions

NATIONAL GEOGRAPHIC: Engineers are always looking for ways to improve solutions to problems. What improvements have you made to the biodigester?

T.H. CULHANE: We have made a couple of improvements so far. We added nets inside the biodigester. Nets increase the surface area on which the bacteria can grow. That means more bacteria and more gas is produced. We also added tubing around the biodigester. We run hot water through these tubes so that the inside of the tank stays warm. Bacteria love warmth. The heat can keep them active and "munching" on our waste in cooler weather.

The tank that holds the gas is installed. It rises and falls with the amount of gas inside the tank.

NG: It sounds like a solar hot water system and the biodigester system can work together. Is that correct?

T.H. CULHANE: Absolutely! Here's an example. I take a shower with solar-heated water. The hot water goes down the drain and into a biodigester. The hot water helps keep the biodigester working well. When the sun isn't shining, the water in the hot water system stays hot in an insulated tank. Suppose that water runs out. Then you have a string of cloudy days. You can use gas from the biodigester to heat more water. So the two systems work together perfectly!

Animal wastes are added to the tank.

Results

NATIONAL GEOGRAPHIC: Engineers analyze their results to see what they have learned and what they might do next. What have you learned from the work your team has done?

T.H. CULHANE: We have learned lots of things. We now know that inexpensive solutions for energy and trash problems really work. We have proven this in places all around the world. Sure, you can buy solar hot water and biodigester systems. But we have shown that you can make these systems yourself. There is very little cost when using recycled materials.

NG: So, what's next?

T.H. CULHANE: Our goal is to help people develop sustainable energy solutions.

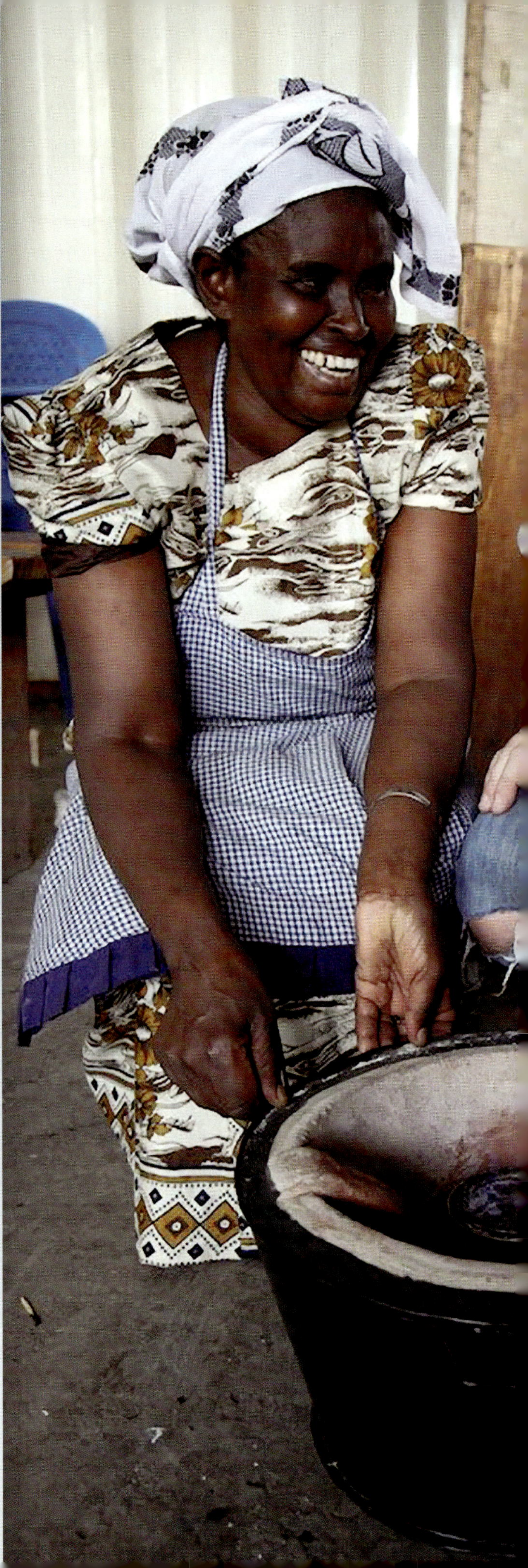

That means using the energy sources available—mostly solar energy. We are teaching people how to use the cheapest solar cells to make their own electricity. We teach people how to build windmills to produce electricity.

Our team is always working with local people to come up with ideas that work. Once you realize there is no such thing as garbage, a whole new world of solutions opens up. Our cities and towns have everything we need to solve our energy problems. Just like the villagers in the rain forest, city dwellers can live sustainably and live well.

"JUST LIKE THE VILLAGERS IN THE RAIN FOREST, CITY DWELLERS CAN LIVE SUSTAINABLY AND LIVE WELL."

T.H. visits a community in Africa. People here have developed this stove. It can get energy from methane gas or from burning wood.

Check In Describe an example in which T.H. Culhane acted like an engineer.

Discuss

1. What connections can you make among the three pieces in *Energy Solutions?*

2. Compare and contrast how people in Nepal get the energy they use and how you get the energy you use at home and school.

3. Why is it difficult to get energy in some places in the world?

4. Describe a time when you have solved a problem by using the approach that engineers use to solve problems.

5. What do you still wonder about solving energy problems? What questions would you ask T.H. Culhane?